I0815322

SNAKES

# BOA CONSTRICTORS

Cody Koala

An Imprint of Pop!

popbooksonline.com

# Hello! My name is Cody Koala

This book is filled with videos, puzzles, games, and more! Scan the QR codes* while you read, or visit the website below to make this book pop.

**popbooksonline.com/boa**

*Scanning QR codes requires a web-enabled smart device with a QR code reader app and a camera.

**abdobooks.com**
Published by Pop!, a division of ABDO, PO Box 398166, Minneapolis, Minnesota 55439. 

Printed in the United States of America, North Mankato, Minnesota.
042025
082025

THIS BOOK CONTAINS RECYCLED MATERIALS

Cover Photo: Shutterstock Images
Interior Photos: Getty Images; Shutterstock Images
Editor: Elizabeth Andrews and Tyler Gieseke
Series Designer: Neil Klinepier, Colleen McLaren

**Library of Congress Control Number: 2024948400**

**Publisher's Cataloging-in-Publication Data**
Names: Murray, Julie, author.
Title: Boa constrictors / by Julie Murray
Description: Minneapolis, Minnesota : Pop!, 2026 | Series: Snakes | Includes online resources and index
Identifiers: ISBN 9781098247799 (lib. bdg.) | ISBN 9781098248338 (ebook)
Subjects: LCSH: Boa constrictor--Juvenile literature. | Pythons--Juvenile literature. | Constrictors (Snakes)--Juvenile literature. | Snakes--Behavior--Juvenile literature. | Snakes--Juvenile literature.
Classification: DDC 597.96--dc23

# Table of Contents

Chapter 1

# Where Do They Live?

Boa constrictors are **reptiles**. They are found in Mexico, Central and South America, and some Caribbean islands.

There are more than 40 true boas that live throughout much of the world.

Boa constrictors live in grasslands, forests, and deserts. They make their homes in logs or empty **dens**. They live alone and only come together to have young.

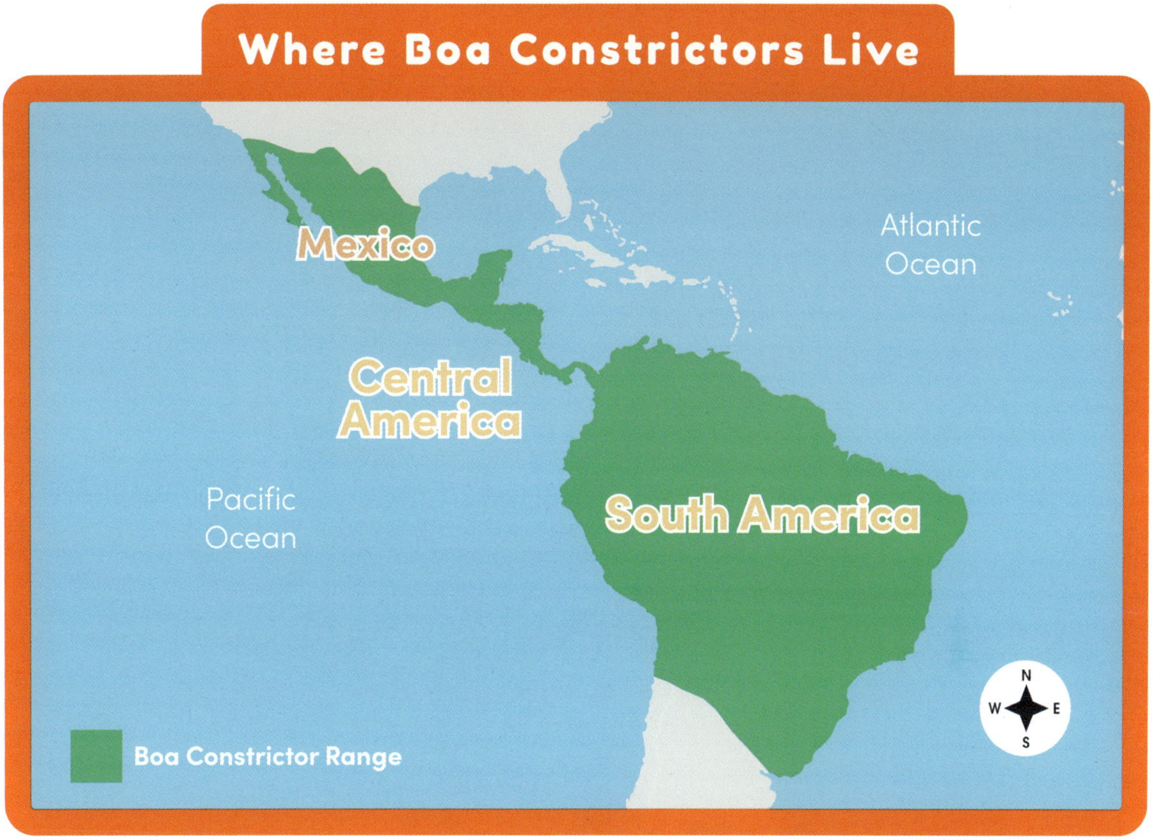

## Where Boa Constrictors Live

Mexico

Central America

South America

Atlantic Ocean

Pacific Ocean

N W E S

Boa Constrictor Range

Boa constrictors move on the ground and in the trees. They often live near

a river or stream. They can swim but like to be on land better.

Chapter 2

# What Do They Look Like?

Boas have heavy bodies covered in **scales**. Most are tan, brown, or cream. They have **saddle**-shaped markings. The markings are light brown on the inside with darker outlines.

Learn more here!

Boa constrictors are big snakes! They can grow 16 feet (4.9m) long. They can weigh more than 100 pounds (45kg).

Boa constrictors shed their skin throughout their life.

A boa constrictor flicks its tongue to pick up scents in the air. It uses its Jacobson's **organ** to smell. This organ is on the roof of its mouth.

Chapter 3

# How Do They Hunt?

Boa constrictors are **ambush** hunters. They strike their **prey** without warning! They grab prey with their sharp teeth. Then, they wrap their body around it and squeeze.

Boa constrictors are nocturnal. This means they move around at night.

Boa constrictors squeeze their prey until it dies. Then, they swallow it whole. They eat birds, mice, and other animals. Adults can eat large animals such as deer!

After a big meal, boa constrictors can go for weeks or months without eating.

Chapter 4

# Boa Babies

Boa constrictors give birth to live babies. They can have up to 65 babies at a time. The babies are 20 inches (50cm) long. They are on their own right away.

Boa constrictors can live for about 30 years in the wild.

**Complete an activity here!**

# Making Connections

## Text-to-Self

Boa constrictors live in many different places. If you were a boa, what kind of place would you want to live in? Why?

## Text-to-Text

Have you read any books about different kinds of snakes? If so, how were they similar to and different from boa constrictors?

## Text-to-World

Boa constrictors are ambush hunters. Can you think of any other animals that hunt this way?

# Glossary

**ambush** – a surprise attack made from a hidden place.

**den** – a safe resting place for wild animals, often underground.

**organ** – a body part that performs a specific function. The heart, lungs, and stomach are organs.

**prey** – an animal that is hunted by other animals for food.

**reptile** – a cold-blooded animal with a skeleton inside its body and dry scales or hard plates on its skin.

**saddle** – a leather seat that is used on the back of a horse to carry a rider.

**scales** – small, hard, thin plates that cover reptiles.

# Index



# Online Resources

popbooksonline.com

Thanks for reading this Cody Koala book!

This book is filled with videos, puzzles, games, and more! Scan the QR codes* while you read, or visit the website below to make this book pop.

popbooksonline.com/boa

*Scanning QR codes requires a web-enabled smart device with a QR code reader app and a camera.